SCHOLASTIC SCIENCE READERS™

FALL LEAVES CHANGE COLORS

KATHLEEN WEIDNER ZOEHFELD

SCHOLASTIC INC.
New York Toronto London Auckland Sydney
Mexico City New Delhi Hong Kong Buenos Aires

PHOTO CREDITS: Cover: Renee Lynn/Photo Researchers. Page 1: Willard Clay/Dembinsky Photo Associates; 3: Dan Dempster/DPA; 4: Richard Phelps Frieman/Photo Researchers; 5, 6: Adam Jones/DPA; 7: Randall B. Henne/DPA; 8: John Kaprielian/Photo Researchers; 9: Mark A. Schneider/DPA; 10: A. B. Sheldon/DPA; 11: David R. Frazier/Photo Researchers; 12–13: Jim Steinberg/Photo Researchers; 14: Willard Clay/DPA; 15: Rod Planck/DPA; 16–17: Eddie Hironaka/Image Bank; 18: Barb Gerlach/DPA; 19: Dan Dempster/DPA; 20: Michael P. Gadomski/Photo Researchers; 21: Mark E. Gibson/DPA; 22, 23: Adam Jones/DPA; 24: Michael Lustbader/Photo Researchers; 25: Jim Zipp/Photo Researchers; 26: Adam Jones/DPA; 27: Michael P. Gadomski/Photo Researchers; 28: Randall B. Henne/DPA; 29: Rod Planck/DPA; 30: Larry Ulrich/Tony Stone.

We are grateful to Francie Alexander, reading specialist, and to Adele M. Brodkin, Ph.D., developmental psychologist, for their contributions to the development of this series.

Our thanks also to our science consultant Jim Sirch,
Director of the Connecticut Audubon Center at Fairfield.

Book design by Barbara Balch and Kay Petronio.
Photo research by Sarah Longacre.

Printed in the U.S.A.

ISBN 0-439-61194-6

1 2 3 4 5 6 7 8 9 10 23 12 11 10 09 08 07 06 05 04 03

It's fall, and you can feel a chill in the air. All around you, the trees are beginning to change.

In your backyard, in the park, and along country roads, trees are putting on their **autumn** (**aw**-tuhm) colors.

Their leaves are turning
yellow, orange, tan, and red.

In summer, most leaves are green. All summer long, something amazing is happening in those green leaves.

Take a Closer Look

A green leaf

They are taking in sunshine, air, and water. And they are making food for the trees!

The green coloring in leaves is called **chlorophyll** (**klor**-uh-fil). Chlorophyll helps leaves take in light from the warm summer sun.

Leaves also have tiny openings for letting in air. From the air, leaves get a special gas called **carbon dioxide** (**car**-bun **deye**-ox-eyed).

After a rain, the trees' **roots** take up water from the ground. Water moves into the leaves.

Using light from the sun, green leaves can turn water and carbon dioxide into sugar.

The sugar moves from the leaves to all parts of the trees. Sugar is the special food that makes trees live and grow.

As summer ends, the days get shorter. The nights get longer. Autumn will soon be here.

Leaves get less and less light each day. This tells the trees that it is time to stop making food. The leaves begin to die.

The chlorophyll in the leaves fades away, and their color changes.

Take a Closer Look

A leaf that is beginning to change color

Each leaf loses its green coloring slowly.

As the chlorophyll fades, other colors in the leaf begin to show.

The leaves of some trees
turn yellow or orange.

These colors are in the leaves all summer. But the strong green of the chlorophyll hides them.

The leaves of other trees turn red or purple.

Sunny days and cool autumn nights make the sugar in the leaves turn red or purple. As the green coloring in the leaves fades, the color of the sugar shows through.

A few types of trees have leaves that turn brown or tan.

In these leaves, a coloring called **tannin** shows as their green coloring fades.

Take a Closer Look

Evergreen needles

Trees whose leaves stay green all year are called evergreens. Most evergreens have tough, narrow leaves called needles.

Evergreens shed their needles only a few at a time.

Trees whose leaves change color lose *all* their leaves in autumn.

As a leaf dies, the part of the stem that holds it to its branch gets weak.

Take a Closer Look

The weakening stem of a leaf

Chilly autumn winds make many leaves fly from their branches. Heavy rains make others fall.

Under your feet, the ground is covered with a patchwork of beautiful colors.

Over your head, the branches look dark and bare. The trees are ready for their winter rest.

Tiny buds line the branches where the leaves once were.

The buds are tough enough to stand the frosty winter.

Take a Closer Look

Buds at the tip of a branch

The buds protect the new green leaves that will unfold in spring.

Glossary

autumn (**aw**-tuhm)—one of the four seasons. Autumn comes after summer and before winter. It is also called fall.

carbon dioxide (**car**-bun **deye**-ox-eyed)—one of the main gases in the air.

chlorophyll (**klor**-uh-fil)—the green coloring found in the leaves of most plants.

roots—the part of a plant that grows under the ground.

tannin—a tan-colored material found in the bark, leaves, roots, and fruits of some plants.

A Note to Parents

Learning to read is such an exciting time in a child's life. You may delight in sharing your favorite fairy tales and picture books with your child.

But don't forget the importance of introducing your child to the world of nonfiction. The ability to read and comprehend factual material will be essential to your child in school, and throughout life. The Scholastic Science Readers™ series was created especially with beginning readers in mind. These books, with their clear texts and beautiful photographs, will help you to share the wonders of science with *your* new reader.

Suggested Activity

Help your child create his or her own autumn leaf album. Go on a walk in a park or wooded area together and collect leaves of every shape, size, and color. Slide your leaves between the pages of a heavy book for a few days to press them. When your leaves are flattened, you can use the Internet or nature guide books to find names for the leaves you've found (oak, birch, sassafras, etc.). With tape or glue, mount your leaves on colorful construction paper and make a label for each. Your child can create a cover for the album with markers or crayons. Using staples or a hole punch and yarn, bind the pieces of paper together to form a book. You've created a reminder of fall that you can enjoy all year round!